# Network Security Essentials

## A Guide to Securing Your Network

## with Firewalla Gold Plus

*An independent, unofficial guide*

L. Phillips

PRAIRIE LIGHT
ESTD. 2025
PRESS

*Helping readers move from the unknown to the known*

# Network Security Essentials

A Guide to Securing Your Network

Third Edition



---

**Disclaimer**

The information in this book is provided for educational and informational purposes only. While every effort has been made to ensure accuracy, technology and security practices evolve over time.

The author makes no guarantees regarding the effectiveness of any tools, techniques, or strategies discussed and assumes no responsibility for errors, omissions, or outcomes resulting from the use of this information.

Readers are encouraged to evaluate their own circumstances and, where appropriate, consult qualified professionals before implementing security, financial, or technical decisions.

---

**Non-Affiliation Notice:** This book is an independent publication and is not affiliated with, endorsed by, or sponsored by any company or organization mentioned herein, including but not limited to Firewalla.

# Table of Contents

# How I Came to Take Home Network Security Seriously

I didn't start thinking seriously about network security because I wanted to become an expert, or because I enjoyed tinkering with technology for its own sake. I started because, over time, I became uncomfortable with how much of my personal life flowed invisibly through my home network—and how little visibility or control I actually had over it.

Like most people, I relied for years on the equipment provided by my internet service provider. It worked well enough. Devices connected. Streaming played. Emails sent. On the surface, everything functioned as expected. But beneath that surface, I realized I had no meaningful way to answer simple questions: What devices were actually on my network? Where was my data going? Which connections were normal, and which weren't?

That discomfort grew as the number of connected devices in my home increased. Phones, laptops, smart TVs, cameras, streaming boxes, voice assistants—each added convenience, but also complexity. Each one introduced software written by someone else, communicating with servers I didn't control, in ways I couldn't easily see. I wasn't paranoid, but I was increasingly aware that "working" wasn't the same as "secure."

I began looking for ways to gain visibility and control without turning my home into an IT project. I wasn't interested in enterprise firewalls, subscription-heavy security services, or tools that required constant attention. I wanted something that respected privacy, ran locally, and made the network understandable rather than opaque.

That search eventually led me to Firewalla.

What stood out wasn't just the feature list, but the philosophy behind it: local processing, clear visibility, and tools that assume the user wants to understand what's happening—not just be told that things are "safe." Over time, as I used Firewalla in my own network, I realized that many of the questions I had struggled to answer were the same questions others were asking as well.

This book grew out of that experience.

It's not written as a marketing guide or a replacement for official documentation. It's written as a practical companion—one that explains not just how to configure Firewalla, but why certain choices matter, how features fit together, and how to think about your home network as a system worth understanding and maintaining.

If you've ever felt that your network works, but you don't really know *how* or *why*, you're in the right place.

# Introduction

Your home network is more than just a way to get online. It's the infrastructure that connects everything in your digital life—computers, phones, smart speakers, security cameras, thermostats, game consoles, streaming devices. Everything.

And like any infrastructure, it needs management, security, and maintenance. That's where Firewalla comes in.

This book is your complete guide to the Firewalla Gold Plus—a powerful network security device that gives you professional-grade protection and control over your home network. Whether you're setting up Firewalla for the first time or looking to master its advanced features, you'll find practical, actionable information here.

## About Firewalla

Firewalla is a hardware firewall and network security device. It sits between your modem and your network, protecting everything behind it while giving you unprecedented visibility and control.

Think of Firewalla as combining two core roles: **Protection & Privacy** and **Control & Management**.

## Protection & Privacy

- **Firewall:** Blocks unwanted connections and intrusion attempts
- **Content filtering:** Blocks ads, trackers, malware, and inappropriate content

- **VPN server:** Lets you securely connect to your network from anywhere
- **VPN client:** Routes traffic through privacy services
- **Network monitoring:** Shows exactly what's happening on your network

### Control & Management

- **Parental controls:** Manages when and what kids can access
- **Network management:** Creates separate networks for different purposes

## Why Do You Need It?

Modern home networks face challenges that didn't exist a decade ago. Consider what's changed:

### More Devices, More Vulnerabilities

The average home now has 25+ connected devices. Each one is a potential entry point for attackers. Your router's basic firewall wasn't designed for this scale or complexity.

### Privacy Is Under Attack

Companies track your browsing across websites. Advertisers build detailed profiles about you and your family. Your ISP can see and log every site you visit. Smart devices send data about your habits back to their manufacturers.

### Kids Need Protection

Children have unprecedented access to the internet through phones, tablets, and computers. Keeping them safe requires more than just hoping they make good choices—it needs technical guardrails.

**IoT Devices Are Insecure**

Smart speakers, cameras, door locks, and thermostats are notoriously poorly secured. Many have known vulnerabilities. Some send data you didn't know they were collecting. They need to be isolated from your important devices.

**Remote Access Needs Security**

Working from home, accessing your NAS while traveling, or checking security cameras remotely all require secure connections. Your router's basic features aren't enough.

**You Want Control**

Your network, your rules. You should decide what's allowed, when it's allowed, and who can do what. Consumer routers offer minimal control. Firewalla gives you complete authority over your network.

## What Makes Firewalla Gold Plus Special?

Firewalla makes several models. The Gold Plus (often called Gold) is the sweet spot for most users—powerful enough for demanding networks but not overkill for typical homes.

### Key Specifications

- Quad-core 1.5 GHz processor: Handles deep packet inspection at full gigabit speeds
- 4 GB RAM: Enough for extensive rule sets and monitoring
- 6 Gigabit Ethernet ports: 1 WAN + 4 LAN + 1 additional multi-purpose port

- Up to 1 Gbps throughput: Won't bottleneck your internet connection
- Up to 5 VLANs: Create virtually separated networks
- Dual-WAN support: Use two internet connections for redundancy or combined bandwidth
- Silent operation: Fanless design, no noise
- Compact size: Fits anywhere, no rack required

## What You Can Do With It

### Security

- Block malware, phishing, and intrusion attempts automatically
- See every connection every device makes
- Isolate vulnerable IoT devices
- Get alerts when suspicious activity occurs
- Block ads and trackers network-wide

### Privacy

- Run your own VPN server for secure remote access
- Route traffic through commercial VPN services
- Block tracking and data collection
- Control what information leaves your network

### Parental Controls

- Filter content by category (adult, violence, gambling, etc.)
- Set time limits on internet usage
- See what sites kids are visiting
- Block specific apps or games
- Create schedules that adapt to school days vs weekends

### Network Management

- Create separate networks (main, guest, IoT, work)

- See bandwidth usage by device
- Prioritize important traffic (video calls over downloads)
- Set bandwidth limits
- Manage two internet connections

## Who This Book Is For

This book is written for people who:

- Just bought a Firewalla Gold Plus and want to set it up properly
- Have Firewalla running but aren't using most of its features
- Want to understand what Firewalla is doing and why
- Need to solve specific problems (slow internet, security concerns, parental controls)
- Want to master advanced features like VLANs and policy-based routing

You don't need to be a network engineer. Technical concepts are explained in plain language with real-world examples. When technical terms are necessary, they're defined clearly.

That said, this isn't a dumbed-down guide. We cover sophisticated features and assume you want to understand how things actually work, not just click buttons. If you want to learn, you'll find detailed explanations. If you just want to get something done, you'll find step-by-step instructions.

## What This Book Covers

The book is organized to work both as a learning path and as a reference:

**Chapters 2-3: Getting Started**

Installation, initial configuration, and basic operation. Get Firewalla protecting your network quickly.

### Chapters 4-7: Core Features

Dashboard navigation, device management, security features, and parental controls. The features you'll use daily.

### Chapters 8-11: Advanced Features

VPN server and client, advanced networking, troubleshooting, and performance monitoring. Power user territory.

### Chapters 12-15: Expert Topics

Advanced rules, security deep-dive, automation, and maintenance. For users who want complete mastery.

### Appendices

Quick reference guides, troubleshooting checklist, and glossary.

## How to Use This Book

### If You're Just Getting Started

Read Chapters 2-7 in order. They build on each other and will take you from unboxing to confident daily use. Don't skip ahead—later chapters assume you understand earlier concepts.

### If You Have Specific Goals

Use the table of contents to find relevant chapters. Want to set up VPN? Chapter 8. Need to create separate networks?

Chapter 9. Troubleshooting issues? Chapter 10. Want to optimize performance? Chapter 11.

### If You're an Experienced User

You can probably skim Chapters 2-7 and focus on Chapters 8-15. But don't assume you know everything in the early chapters—Firewalla has some non-obvious features even experienced users miss.

### As a Reference

Keep this book handy. When you encounter a feature or setting you don't understand, look it up. The index and table of contents make it easy to find information quickly.

## Conventions Used in This Book

### Navigation Paths

When instructions say to tap or click through menus, they're written like this: Settings -> System -> Updates. This means tap Settings, then System, then Updates.

### Important Notes

Key information is called out in context. Pay attention to paragraphs that start with words like Important, Note, or Warning.

### Screenshots

The book includes screenshots showing what you should see in the app. Your screen might look slightly different depending on your Firewalla model, app version, or network configuration, but the basic layout will be the same.

### Example Scenarios

Throughout the book, you'll find example scenarios showing how to use features in real situations. These aren't hypothetical—they're based on actual use cases from real Firewalla users.

## What You Need

### Hardware

- Firewalla Gold Plus
- Internet connection (cable, fiber, DSL, etc.)
- Modem or ISP-provided gateway
- Smartphone or tablet for initial setup (iOS or Android)
- Ethernet cables

### Optional But Recommended

- Managed network switch if you plan to use VLANs
- Separate Wi-Fi access point or mesh system (Firewalla doesn't include Wi-Fi)
- Computer for accessing web interface (some features are web-only)
- Backup internet connection if you're setting up dual-WAN

### Knowledge

No networking expertise required, but basic computer skills help. You should be comfortable installing apps on your phone, connecting devices with Ethernet cables, following step-by-step instructions, and using a web browser.

Everything else you need to know is explained in the book.

## Getting Help

This book covers most situations, but you might still have questions. Here's where to find help:

**Firewalla Support**

- Help center: help.firewalla.com - Searchable knowledge base
- Community forum: Discussions with other users and Firewalla staff
- Email support: help@firewalla.com
- In-app support: Tap the question mark icon in the app

**This Book**

- Index: Find specific topics quickly
- Troubleshooting appendix: Common problems and solutions
- Glossary: Definitions of technical terms

## A Note About Versions

Firewalla releases updates regularly. The screenshots and specific menu locations in this book are accurate as of December 2024. If your app looks slightly different, that's normal—Firewalla improves the interface over time.

The core concepts and features remain stable. Even if a button moves or a screen is redesigned, the underlying functionality works the same way. Focus on understanding what features do, not just where to click.

## Privacy and Data Collection

Firewalla takes privacy seriously. Understanding what data it collects and where it goes matters.

**Local Processing**

All security features, rule enforcement, and monitoring happen on your Firewalla box. Your network traffic doesn't route through Firewalla's cloud servers. They can't see what you're doing because the data never reaches them.

**What Firewalla Cloud Receives**

- Device names and groups (for multi-device app sync)
- Rule configurations (so they sync across devices)
- Anonymous usage statistics (what features are used, not what traffic flows through)
- Alarm summaries for notifications (but not full traffic logs)

**What Firewalla Never Receives**

- Your actual internet traffic content
- Websites you visit
- Files you download
- Messages or communications
- Video streams or media content

You control your data. If you prefer, you can disable cloud sync entirely and manage everything locally, though you'll lose convenient features like remote access to the app and push notifications.

## Ready to Begin

Network security isn't just for IT professionals anymore. Your home network deserves the same protection businesses get. Firewalla brings enterprise-grade security to regular people, with an interface that makes sense and features you can actually use.

This book will teach you to use Firewalla effectively. You'll understand not just what to click, but why you're clicking it. You'll know how to diagnose problems, optimize performance, and configure advanced features with confidence.

By the time you finish, you'll have a secure, well-managed network that protects your family, respects your privacy, and works the way you want it to.

Ready? The next chapter covers getting Firewalla installed and configured.

# Chapter 1: Meet Your Firewalla Gold Plus

Before we dive into installation and configuration, let's get acquainted with what you've purchased. The Firewalla Gold Plus isn't just another router or security appliance—it's a complete network security platform that happens to fit in the palm of your hand.

Understanding what your Firewalla Gold Plus can do, and equally important, understanding its design philosophy, will help you make the most of its capabilities. This chapter walks you through the hardware, the core features, and the thinking behind why it works the way it does.

## The Hardware

### Physical Design

The Firewalla Gold Plus is compact—roughly the size of a small hardcover book. The metal chassis serves dual purposes: it protects the internal components and acts as a heat sink for passive cooling. There's no fan, which means no noise and one less component to fail.

The fanless design isn't just about silence. Fans collect dust, wear out bearings, and become points of failure. A passively cooled system that's properly designed will outlast its actively cooled counterparts. The chassis gets warm during operation—that's normal and by design. It's dissipating heat effectively without requiring moving parts.

The unit weighs enough to feel substantial but not so much that you can't mount it on a wall. Speaking of which, it comes with mounting hardware if you want to keep it out of the way. Most people either mount it near their networking equipment or simply set it on a shelf where it has good ventilation.

## Ports and Connections

All the action happens on the back panel:

- **Four 2.5 Gigabit Ethernet ports: These are your network connections. One typically connects to your internet (WAN), and the others connect to your local network (LAN). But Firewalla is flexible—you can configure these ports in various ways depending on your needs.** They're color-coded and numbered, which helps during setup.
- **Two USB 3.0 ports:** One is occupied by a red security dongle (don't remove this—it's tied to your device). The other can be used for the optional Wi-Fi adapter or future expansions.
- **HDMI port:** This provides console access if you ever need to troubleshoot at a low level. Most users will never touch this.
- **Serial console port:** Another troubleshooting interface that you'll likely never need. It's there for advanced scenarios.
- **Power input:** Standard DC power input. The included power adapter is compact and efficient, drawing less than 15 watts on average—about the same as an LED light bulb.

The front panel has a power button (primarily for shutdown) and a reset button. The reset button should only be used as a last resort when something goes wrong. The power button is mainly for graceful shutdowns—the system boots automatically when power is applied.

## The 2.5 Gigabit Advantage

Most home networking equipment uses Gigabit Ethernet (1 Gbps). The Gold Plus uses 2.5 Gigabit (2.5 Gbps) ports. This might seem like overkill if your internet connection is "only" 1 Gbps, but there's method to this apparent madness.

First, internet speeds are increasing. Multi-gigabit fiber is becoming more common, and cable companies are pushing speeds beyond 1 Gbps. The Gold Plus is ready for this without requiring replacement.

Second, and more immediately relevant: security processing has overhead. When Firewalla examines every packet flowing through your network—checking for threats, applying rules, logging activity—that takes processing power. With 1 Gbps ports, you might see your effective throughput drop to 700-800 Mbps under heavy security processing. With 2.5 Gbps ports, you maintain full gigabit speeds even with comprehensive security enabled.

Third, your internal network traffic benefits. Backing up a computer to a NAS, streaming high-bitrate video from a local server, or transferring large files between devices—these activities happen at LAN speeds, not internet speeds. The faster ports mean these local transfers complete faster.

The Gold Plus is rated for up to 5 Gbps of total packet processing throughput. This means it can handle multiple 2.5 Gbps streams simultaneously or process data at line rate with security features enabled.

## Inside the Box

The Gold Plus is built around an Intel J4125 processor—a quad-core chip designed for efficiency and reliability. It has 4GB of RAM and runs a custom Linux distribution that's been optimized for network security tasks.

This is important because unlike consumer routers that use generic firmware, Firewalla's software is purpose-built. Every component of the system is designed to work together: the hardware acceleration for encryption, the network processing, the storage management, and the security analysis.

The system boots from eMMC flash storage—solid-state memory that's faster and more reliable than SD cards or spinning hard drives. This contributes to the quick boot times (the system is typically ready in under a minute) and the system's overall responsiveness.

## Core Capabilities

The hardware is the foundation, but the software is where Firewalla really shines. Let's walk through what it actually does for you.

### Network Visibility

The first thing Firewalla does is show you what's happening on your network. Not in the abstract "there's traffic" sense, but specifically: what device is connecting to what service, how much bandwidth it's using, when it's active, and what kind of data is flowing.

Every device on your network gets identified. Firewalla recognizes the manufacturer, device type, and in many cases, the specific model. Your iPhone shows up as an iPhone, not just as "Unknown Device." Your smart TV is labeled as a TV. Your printer is clearly marked as a printer.

For each device, you can see:

- **What domains and services it connects to:** Is your smart TV talking to Netflix, or is it also sending data to advertising networks?
- **Bandwidth usage over time:** Which devices are consuming the most data, and when?
- **Connection patterns:** When is the device most active? Are there unusual connection spikes?
- **Security alerts:** Any blocked threats or suspicious activity associated with this device.

This visibility isn't about surveillance—it's about understanding. When something goes wrong (your internet seems slow, a device isn't connecting properly, you're worried about security), you have the data to figure out what's actually happening rather than guessing.

### Active Protection

Firewalla doesn't just observe—it protects. The system includes several layers of active defense that work automatically once enabled:

### Intrusion Detection and Prevention (IDS/IPS)

The system watches for known attack patterns and malicious behavior. When it detects something suspicious—port scanning, exploit attempts, command-and-control traffic—it can alert you or block it automatically.

This happens at network speed. The IDS/IPS engine examines traffic in real-time, comparing it against a database of threat signatures that's updated regularly. When you enable Active Protect, you're not just blocking known malware sites—you're stopping exploit attempts, preventing unauthorized access, and catching threats before they reach your devices.

### Ad and Tracker Blocking

Firewalla can block ads and trackers at the network level. This is different from browser ad blockers—it works for every device and every app on your network. Your smart TV's ads disappear. Your kids' tablets stop showing inappropriate advertising. Mobile apps that are riddled with ads become usable.

The ad blocking uses curated blocklists that identify advertising and tracking domains. When a device tries to connect to these domains, Firewalla simply refuses the connection. To the app or website, it looks like the ad server is unavailable. The content loads, minus the ads.

You can customize this extensively. Use aggressive blocking to remove nearly all ads and trackers, or use lighter filtering that only blocks the most intrusive content. You can also whitelist specific sites if blocking causes problems.

### Malware and Phishing Protection

The system maintains blocklists of known malware distribution sites, phishing domains, and command-and-control servers. These are updated continuously. When any device on your network tries to connect to a blocked domain, the connection is refused and you receive an alert.

This protection works even if the device itself is compromised. A infected computer trying to "phone home" to its controllers will be blocked. A user clicking a phishing link will be stopped before reaching the fake site. This is your network-level immune system.

### Behavioral Analysis

Beyond signature-based detection, Firewalla watches for unusual behavior. A device that suddenly starts uploading massive amounts of data, a new device appearing on your network without your knowledge, connection patterns that don't match normal usage—these trigger alerts.

The system learns what's normal for your network over time. This means it gets better at distinguishing between legitimate activity and potential problems. It's not about paranoid false alarms—it's about surfacing genuinely unusual events that deserve your attention.

### Network Control

Visibility and protection are important, but control is where you shape your network to work the way you want:

### Rules and Policies

You can create rules that control what happens on your network. Block a specific website. Limit bandwidth for a device or application. Prevent certain devices from communicating with each other. Route traffic through a VPN. The possibilities are extensive.

Rules can be simple ("Block Facebook on my kids' devices") or sophisticated ("Route all traffic from my work laptop through my company VPN, except local network traffic, but only during business hours"). The app provides templates for common scenarios, so you don't need to think like a network engineer to create useful policies.

### Family Protect Controls

Firewalla includes robust parental control features. You can filter content by category (block adult content, gambling, social media, gaming), set usage schedules (internet access only during certain hours), and pause the internet for specific devices or users.

The "Social Hour" feature is particularly clever—it blocks all social media for one-hour increments, making it easy to enforce screen-free family time without completely cutting off internet access. Kids can still do homework that requires web access; they just can't scroll TikTok during dinner.

Safe Search enforcement ensures that search engines return filtered results, even if the child tries to disable safe search. And all of this works regardless of what device they're using or whether they've installed blocking software—it's enforced at the network level.

### Network Segmentation

One of the Gold Plus's most powerful features is the ability to create separate network segments. Think of these as virtual walls within your network.

You might create a guest network that has internet access but can't see or connect to your personal devices. Or an IoT network for smart home devices that's isolated from your computers and phones. Or a kids' network with strict filtering and time limits while your main network remains unrestricted.

Each segment can have its own rules, policies, and security settings. Devices on one segment can't see devices on another unless you explicitly allow it. This is how enterprises protect their networks, and now you can do the same at home.

### Bandwidth Management

Smart Queue is Firewalla's traffic prioritization feature. It ensures that your video calls don't lag when someone else is downloading a large file. Gaming traffic gets priority when needed. Video streaming is managed to prevent buffering without hogging all the bandwidth.

You can also set hard limits: this device gets no more than X Mbps, or this activity type gets throttled during certain hours. It's your bandwidth—you decide how it's allocated.

### VPN Capabilities

The Gold Plus includes comprehensive VPN support in both directions:

### VPN Server

When you're away from home, you can connect back to your home network securely. This means:

- Access to your home network resources (files, printers, security cameras) from anywhere
- Your internet traffic appears to come from home, which can be useful for accessing geo-restricted content or avoiding suspicious public Wi-Fi networks
- All the security and filtering you have at home extends to your devices wherever you are

The system supports both OpenVPN and WireGuard protocols. WireGuard is newer, faster, and particularly good for mobile devices. OpenVPN is more established and works through more types of networks. You can enable both simultaneously.

Setting up VPN access is remarkably simple—the app generates configuration profiles that you can install on your devices with a few taps. No manual configuration of IP addresses or certificates required.

### VPN Client

Firewalla can also act as a VPN client, connecting your entire network (or specific devices) through a commercial VPN service. This is useful if you want to:

- Route all your home traffic through a VPN for privacy
- Access content as if you're in a different country
- Protect devices that can't run VPN software themselves (smart TVs, game consoles)

The advantage of doing VPN at the router level is that it covers every device automatically. You don't need to configure VPN on each computer, phone, and tablet. Change VPN providers or servers? Update it once in Firewalla and all your devices get the new settings.

You can even get selective: route only certain devices through the VPN, or only certain types of traffic. Your work laptop goes through your company VPN while everything else uses your regular internet connection. Your smart TV goes through a VPN for streaming while your gaming console uses the direct connection for lower latency.

## Router Mode vs. Simple Mode

The Firewalla Gold Plus can operate in different modes, and understanding these modes will help you set it up correctly for your situation.

### Simple Mode (DHCP Mode)

In Simple Mode, Firewalla sits alongside your existing router. Your router continues to handle routing and DHCP (assigning IP addresses to devices). Firewalla monitors all the traffic and applies its security features, but it doesn't replace your router.

This mode is ideal when:

- You rent your router from your ISP and can't easily replace it
- You have a mesh network system that you want to keep using
- You want to add security without changing your current network setup
- You're not comfortable with advanced networking changes

The installation is straightforward: plug Firewalla into one of your router's LAN ports and connect your devices to Firewalla's ports. The system automatically configures itself.

Simple Mode gives you all of Firewalla's security and monitoring features. The main limitation is that you can't take full advantage of network segmentation or advanced routing features—your existing router is still handling those functions.

### Router Mode

In Router Mode, Firewalla becomes your router. It connects directly to your modem and handles all routing, DHCP, and network management duties. Your existing router either gets removed or is put into "bridge mode" (effectively turning it into just a Wi-Fi access point).

This mode unlocks the full capabilities of the Gold Plus:

- Full network segmentation with isolated VLANs
- Multi-WAN support (multiple internet connections, load balancing, and failover)
- Advanced routing policies
- Complete control over DHCP and DNS settings
- Better performance (no double-NAT or routing overhead)

Router Mode is the recommended configuration if you own your equipment and are comfortable making networking changes. It's not particularly difficult—the app walks you through it—but it does require connecting to your modem directly and potentially reconfiguring your Wi-Fi access points.

One important note: the Gold Plus doesn't have built-in Wi-Fi. In Router Mode, you'll need separate Wi-Fi access points or routers running in access point mode. This might sound like a limitation, but it's actually an advantage—you can use whichever Wi-Fi solution works best for your space, and you can upgrade your Wi-Fi independently from your security and routing.

### Bridge Mode

Bridge Mode is a specialized configuration where Firewalla sits transparently between your network segments. It's used in specific scenarios where you need to monitor traffic without changing IP addresses or network structure—typically for compatibility with complex existing setups.

Most users won't need Bridge Mode. It's mentioned here for completeness, but Simple Mode or Router Mode will handle the vast majority of home and small office setups.

## What You Can't Do (And Why That's OK)

No product does everything, and Firewalla makes specific design choices that mean certain things aren't possible or aren't prioritized. Understanding these limitations helps set appropriate expectations.

### No Built-in Wi-Fi

The Gold Plus doesn't include Wi-Fi. This is intentional—Wi-Fi access points and security appliances are different jobs that require different hardware. By separating them, you get better performance for both functions and the flexibility to choose the Wi-Fi solution that works for your space.

You can add a Wi-Fi USB adapter for specific use cases (like creating a temporary network or using a phone hotspot as backup internet), but for your main Wi-Fi, you'll use separate access points.

### Not a Next-Gen Firewall for Enterprises

While the Gold Plus is powerful, it's designed for homes and small offices, not enterprise deployments. It doesn't have features like integration with Active Directory, sophisticated compliance reporting, or the redundancy and failover capabilities that large organizations require.

For a home or small business, this is perfectly appropriate. You get professional-grade security without paying for enterprise features you don't need.

### No Subscription Required (But No Subscription Available Either)

Firewalla doesn't charge monthly or annual fees. You buy the hardware, and all features and updates are included. This is great for cost predictability, but it means certain service-based features aren't available.

For example, you can't pay for Firewalla to provide 24/7 monitoring and incident response. You're operating your own security—Firewalla gives you the tools, but you're the one using them. For most people, this trade-off (no recurring costs, but you're responsible for responding to alerts) is acceptable.

## The Mobile App Philosophy

One of Firewalla's defining characteristics is that it's managed entirely through a mobile app. There's a web interface (currently in beta), but the app is the primary interface by design.

This might seem unusual if you're accustomed to configuring network equipment through web browsers. But consider: when was the last time you wanted to tweak your network security while sitting at a desk? Network security isn't something you configure once and forget—it's something you check from the couch, adjust when you see an alert on your phone, and modify when your kids ask why they can't access a site.

The app design reflects this usage pattern. It's information-dense but still clean. It provides quick access to common tasks (block this device, pause the internet, check bandwidth usage) while keeping advanced features available but out of the way.

Notifications appear where notifications belong—on your phone's lock screen. You don't need to remember to check a dashboard; alerts come to you. See something suspicious at 2 AM? Address it from your phone without getting out of bed.

Multiple people can have the app connected to the same Firewalla box. This is useful for families where parents might want to manage controls separately, or small businesses where multiple people share administrative duties.

## Updates and Improvement

Firewalla's software is under continuous development. Updates are frequent—typically every few weeks—and they include both security updates and new features.

Updates happen automatically unless you disable this (not recommended). The system downloads updates in the background and applies them during low-traffic periods. You'll get a notification that an update was installed, but there's typically no disruption to your network.

This matters because the threat landscape changes constantly. New vulnerabilities are discovered, new attack patterns emerge, and new types of malicious behavior develop. Firewalla's update mechanism means you benefit from these improvements without having to do anything.

The company also maintains an active community forum where users discuss configurations, share tips, and request features. Many of the features added over the years came from user suggestions. This isn't abandoned-after-purchase hardware—it's an actively supported and improving platform.

## Knowing Your Equipment

By now you should have a solid understanding of what the Firewalla Gold Plus is: a network security appliance designed for homes and small businesses that's powerful enough to provide professional-grade protection but accessible enough that you don't need to be a network engineer to use it effectively.

You know the hardware—four 2.5 Gbps ports, fanless cooling, efficient power consumption. You understand the core capabilities—visibility, protection, control, and VPN. You've learned about the different operating modes and when to use each. And you understand the design philosophy: mobile-first interface, continuous updates, and features that prioritize practical protection over theoretical completeness.

In the next chapter, we'll move from theory to practice and get your Firewalla Gold Plus installed and running on your network.

# Chapter 2: Installation and Initial Setup

Installing your Firewalla Gold Plus is straightforward—most people complete the process in under an hour. This chapter walks you through everything from choosing your installation mode to verifying that your system is working correctly.

We'll take this step by step, addressing common questions and potential issues along the way. By the end of this chapter, your network will be protected and you'll understand how everything connects together.

## Before You Begin

### What You Need

Gather these items before starting:

- **Your Firewalla Gold Plus:** Make sure the red security dongle is plugged into one of the USB ports. Don't remove this—it's tied to your specific device.
- **Power adapter:** The included power supply. Use only the supplied adapter—using incorrect power supplies can damage the unit.
- **Ethernet cables:** You'll need at least two. One is typically included. Cat5e or Cat6 cables work fine—you don't need special cables for 2.5 Gbps; standard Gigabit Ethernet cables handle it.
- **Your smartphone:** iOS or Android. You'll need the Firewalla app, which is free in the App Store or Google Play.
- **Network information:** If you have custom network settings (static IPs, specific DNS servers, VLANs), have this information handy. Most people won't need this.

- **30-60 minutes:** Allocate uninterrupted time. Internet access will be briefly disrupted during setup, so don't start this right before an important video call.

## Choosing Your Installation Mode

Before plugging anything in, decide which mode you'll use. This affects how you connect the cables. The decision tree is simple:

**Choose Simple Mode If:**

- You rent your modem/router combo from your ISP
- You have a mesh Wi-Fi system (Google Wifi, Eero, etc.) that you want to keep using as-is
- You're not comfortable changing your network's current structure
- You want the simplest possible installation

**Choose Router Mode If:**

- You own your equipment
- You want maximum control and features
- You're comfortable putting your existing router in bridge mode or replacing it
- You need network segmentation or multi-WAN features

When in doubt, start with Simple Mode. You can always switch to Router Mode later—it's not a permanent decision. Simple Mode gets you protected quickly and lets you learn the system before making more significant network changes.

## Understanding Your Current Setup

Take a moment to understand your current network. This helps you know where Firewalla will fit:

**Modem Only**

If you have a separate modem (cable modem, fiber ONT, or DSL modem) and a separate router, your setup is straightforward. The modem connects to the internet, and the router provides Wi-Fi and local network services. For Router Mode, Firewalla goes between these. For Simple Mode, it connects to your router.

### Gateway (Combined Modem/Router)

Many ISPs provide a gateway—a single device that's both modem and router. These often have Wi-Fi built in. If you have one of these, Simple Mode is usually the easiest path. You can use Router Mode, but it requires putting the gateway into "bridge mode" or "passthrough mode," which varies by manufacturer and isn't always well-documented.

### Mesh Systems

If you have a mesh Wi-Fi system like Google Wifi, Eero, or similar, you can either use Simple Mode (Firewalla connects to your mesh system's primary node) or Router Mode (mesh system goes into bridge/access point mode, Firewalla becomes the router). Both work fine; it's a matter of preference.

# Simple Mode Installation

Simple Mode installation puts Firewalla behind your existing router. Your router continues to assign IP addresses and handle routing; Firewalla monitors and protects the traffic.

## Physical Connections

### Step 1: Prepare Your Router

Locate an available LAN port on your router. This is any port that's not labeled "WAN" or "Internet." If all your LAN ports are occupied, you may need to temporarily unplug something or add a switch.

### Step 2: Connect Firewalla to Router

Using an Ethernet cable, connect Port 4 on the Firewalla (the port furthest from the power input) to a LAN port on your router. Port 4 is typically labeled or marked with a different color.

### Step 3: Connect Your Devices to Firewalla

Connect your devices to the remaining ports on Firewalla (Ports 1, 2, and 3). You can connect:

- Individual devices (computers, game consoles, smart TVs)
- A network switch, which then connects to multiple devices
- Wi-Fi access points

Not all ports need to be used. If you only have one or two wired devices, that's fine.

### Step 4: Connect Power

Plug in the power adapter. The Gold Plus will power on automatically—there's no power switch. You'll hear a beep, and the unit will begin booting. Wait about 30-60 seconds for the boot process to complete.

## App Pairing and Configuration

### Step 1: Install the App

Download the Firewalla app from the App Store (iOS) or Google Play (Android). Open the app and create an account if you don't already have one. This account is used to sync your settings and access your Firewalla remotely—it's free and doesn't require any payment information.

### Step 2: Start Pairing

Make sure your phone is connected to your Wi-Fi network (the same network that your router provides). Tap the "+"

button in the app and select "Firewalla Gold" or "Firewalla Gold Plus."

The app will search for your Firewalla on the network. This usually takes 10-30 seconds. If it doesn't find your Firewalla, make sure your phone is on the same network and that the Firewalla has finished booting (you should see lights on the Ethernet ports).

**Step 3: Scan QR Code**

When prompted, scan the QR code on the bottom of your Firewalla Gold Plus. The app uses your phone's camera for this. The QR code uniquely identifies your device and pairs it with your account.

Note: On Android devices, the app might accidentally scan the serial number next to the QR code instead of the QR code itself. If this happens, use your finger to cover the serial number text and scan again.

**Step 4: Select Simple Mode**

The app will ask you to select an installation mode. Choose "Simple Mode" or "DHCP Mode" (these are the same thing—the app uses both terms). Confirm your wiring matches what the app shows: Port 4 to router, other ports to devices.

**Step 5: Automatic Configuration**

Firewalla will automatically configure itself. This takes 2-5 minutes. During this time:

- The system configures network settings
- Downloads any pending software updates
- Initializes security databases
- Begins discovering devices on your network

Your internet access shouldn't be interrupted during this initial configuration.

### Step 6: Name Your Box

The app will ask you to name your Firewalla box. This name is only for your reference—it doesn't affect anything technical. Use something that makes sense for your location ("Home Firewalla" or "Office" or "Main House").

## Initial Configuration Wizard

After the basic setup completes, Firewalla presents a configuration wizard that helps you enable core features. You can skip this and configure everything later, but it's worth going through—it only takes a few minutes and gets you protected immediately.

### Enable Ad Blocking

The wizard will ask if you want to enable ad blocking. Say yes. You can choose the level:

- **Strict:** Blocks most ads and trackers aggressively. May occasionally block legitimate content.
- **Normal:** Balanced blocking. Removes most ads while minimizing false positives.
- **Light:** Only blocks the most intrusive ads and trackers.

Start with Normal. You can adjust this later if you find it's too aggressive or not aggressive enough.

### Enable Security Features

The wizard will recommend enabling:

- **Malware Blocking:** Prevents connections to known malicious domains. Enable this.
- **Phishing Protection:** Blocks known phishing sites. Enable this.
- **Active Protect (IPS):** Intrusion prevention that blocks exploit attempts. This is more intensive and can

occasionally cause compatibility issues with certain applications. You can enable it now or later after you've familiarized yourself with the system.

For the first-time setup, enable malware and phishing protection. Consider enabling Active Protect after a week or two when you're comfortable with the system and can recognize whether it's causing any problems.

**Family and Time Controls**

If you have children, the wizard will help you set up parental controls:

- **Safe Search:** Forces search engines to filter results
- **Content Filtering:** Blocks categories like adult content, gambling, etc.
- **Time Schedules:** Sets when internet access is available

You can configure these now or skip and set them up later in the Family section of the app.

## Router Mode Installation

Router Mode installation makes Firewalla your primary router. This requires more steps than Simple Mode but unlocks all capabilities.

### Prerequisites

Before starting Router Mode setup:

- **Know your modem's location:** You'll be connecting directly to it.
- **Plan for Wi-Fi:** Firewalla doesn't have Wi-Fi. You'll need access points or a router in bridge mode.
- **Prepare for brief downtime:** Your internet will be down for 5-10 minutes during the transition.

- **Reboot your modem:** Some ISPs require this when changing routers. Power cycle your modem before connecting Firewalla.

## Physical Connections

### Step 1: Connect Firewalla to Modem

Using an Ethernet cable, connect Port 4 on the Firewalla to the LAN/Ethernet port on your modem. If you have a combined modem/router (gateway), you might need to enable bridge mode on it first—check your ISP's documentation for how to do this.

### Step 2: Connect Wi-Fi Access Points

Connect your Wi-Fi access points or router (in bridge/AP mode) to one of Firewalla's LAN ports (1, 2, or 3). If you have multiple access points or need more ports, connect a switch to one of these ports.

### Step 3: Connect Wired Devices

Connect any wired devices to the remaining LAN ports or to a switch that's connected to Firewalla.

### Step 4: Power On

Connect the power adapter. Wait 30-60 seconds for boot.

## App Pairing and Configuration

### Step 1: Special Connectivity Note

During Router Mode setup, your phone needs to pair with Firewalla before you have internet access. This means:

- If using the app on the same phone, temporarily switch to cellular data during setup
- Or use a different phone that still has internet access

The app will guide you through this.

**Step 2: Start Pairing**

Open the Firewalla app, tap "+", and select Firewalla Gold/Gold Plus. The app will discover your device via Bluetooth and the local network.

**Step 3: Scan QR Code**

Scan the QR code on the bottom of your Gold Plus when prompted.

**Step 4: Select Router Mode**

Choose "Router Mode" when asked about installation type. Confirm your wiring: Port 4 to modem, other ports to your network.

**Step 5: Network Configuration**

Firewalla will:

- Attempt to get an IP address from your modem (WAN IP)
- Configure itself as a DHCP server for your LAN
- Set up DNS services
- Enable routing between WAN and LAN

This takes 5-10 minutes. Your devices will lose internet access during this time as they transition from your old router to Firewalla.

**Step 6: Device Reconnection**

After Firewalla completes configuration:

- Wi-Fi devices: Reconnect to your Wi-Fi. The network name should be the same if you kept your access points, but devices will get new IP addresses.
- Wired devices: Should reconnect automatically.

Your phone may need to be reconnected to Wi-Fi to continue with the app.

### Router Mode Network Settings

In Router Mode, Firewalla gives you control over network settings:

#### IP Address Range

By default, Firewalla uses the 192.168.218.x range for your LAN. This is chosen to avoid conflicts with common router defaults (192.168.1.x or 192.168.0.x). You can change this in Network Manager if you need a different range.

#### DHCP Settings

Firewalla automatically assigns IP addresses to devices. You can:

- Reserve specific IPs for devices (useful for servers or printers)
- Adjust the DHCP range
- Set lease times

#### DNS Settings

By default, Firewalla uses its own DNS service, which enables ad blocking and security filtering. You can change to:

- Your ISP's DNS
- Public DNS services (Google 8.8.8.8, Cloudflare 1.1.1.1, etc.)
- DNS over HTTPS (DoH) for encrypted DNS queries

Stick with Firewalla's DNS initially—it's required for many security features to work.

## Verifying Your Installation

Once setup completes, verify everything is working correctly.

#### Check Internet Access

The most basic check: open a web browser and visit a few websites. If you can access the internet, the fundamental setup is working. Try:

- A major news site (CNN, BBC)
- A streaming service you use
- A search engine

If nothing loads, check the troubleshooting section below.

### Review Device List

In the Firewalla app, go to the Devices section. You should see a list of devices on your network. Firewalla discovers devices over time—you might not see everything immediately, especially devices that are currently sleeping or powered off.

For each device:

- Check that it's identified correctly (or give it a meaningful name)
- Verify it's online if it should be
- Look at what services it's connecting to

This is also a good time to identify any unknown devices. If you see devices you don't recognize, they might be:

- Smart home devices you forgot about
- Neighbor devices that were previously connected (if you kept the same Wi-Fi password)
- Legitimate but poorly-identified devices

You can block or quarantine unknown devices until you identify them.

## Test Security Features

### Ad Blocking Test

Visit a website known for heavy advertising. Compare the experience to how it looks on a device not going through Firewalla (like your phone on cellular). You should see significantly fewer ads.

Check the Firewalla app's main screen—it shows blocking statistics. You should see numbers under "Ads Blocked" and "Trackers Blocked" increasing as you browse.

### Security Test

Don't try to visit actually malicious sites to test protection—that's unsafe even with protection enabled. Instead, check that the security features are active:

- In the app, go to Features → Security
- Verify that Malware Blocking and Phishing Protection show as "On"
- Check if Active Protect (IPS) is enabled if you chose to turn it on

## Common Setup Issues

### No Internet Access

**In Simple Mode:**

- Verify Port 4 is connected to your router's LAN port (not WAN)
- Make sure your router's DHCP is enabled
- Try rebooting both your router and Firewalla

**In Router Mode:**

- Verify Port 4 is connected to your modem
- Power cycle your modem (some ISPs require this when changing routers)

- Check if your ISP requires MAC address registration (some cable ISPs do)
- If you have a gateway, ensure it's in bridge mode

## App Can't Find Firewalla

- Ensure your phone is on the same network as Firewalla

- Enable Bluetooth on your phone (Firewalla uses Bluetooth for initial discovery)
- Wait longer—discovery can take up to a minute
- Try moving your phone closer to the Firewalla box
- Restart the Firewalla app

## Devices Not Getting IP Addresses (Router Mode)

- Verify Firewalla completed its setup (check the app for confirmation)
- Make sure you're not still running DHCP on your old router
- Try manually releasing and renewing IP addresses on problem devices
- Check Network Manager in the app to verify DHCP is enabled

## Slow Speeds

If your internet feels slower after installing Firewalla:

- Run a speed test to establish baseline performance
- Check if Active Protect (IPS) is enabled—it's intensive and might impact speeds on very fast connections
- Verify you're using Cat5e or better Ethernet cables
- Check if Smart Queue is enabled and configured correctly

The Gold Plus should handle gigabit speeds easily. If you're experiencing significant slowdowns, something is misconfigured.

## Next Steps

Congratulations—your Firewalla Gold Plus is installed and protecting your network. Take some time to explore the app and get comfortable with the interface.

In the next chapter, we'll walk through the dashboard and help you understand what you're seeing. You'll learn to interpret the graphs, understand the alerts, and find the information you need quickly.

For now, let Firewalla run for a day or two. It will learn your network's normal patterns, discover all your devices, and start showing you insights about your traffic. The more it runs, the better it gets at distinguishing normal activity from things that deserve your attention.

# Chapter 3: Understanding Your Network Dashboard

Your Firewalla dashboard is where you'll spend most of your time. It's the command center for your network—showing you what's happening right now, what happened yesterday, and what needs your attention.

This chapter walks through every section of the dashboard, explaining what you're seeing and why it matters. By the end, you'll be able to glance at your phone and immediately understand your network's status.

### Network Statistics

Below the graph, you'll see key statistics about your network. These numbers update continuously and provide quick answers to common questions.

#### Devices Online

This shows how many devices are currently connected to your network. The number includes everything: computers, phones, tablets, smart home devices, game consoles, printers—anything with an IP address.

## Quick Access Buttons

Below the statistics, you'll find large buttons for the most frequently accessed features. These provide one-tap access to common tasks.

### Devices

Tap this to see your complete device list. This is where you manage everything connected to your network.

### Alarms

This opens your alarm history. Alarms are sorted by time, with the most recent at the top.

*The Alarms page shows all network events that require attention.*

## Common Dashboard Workflows

Here are typical scenarios and how to use the dashboard to handle them.

### Checking Network Health

When you want to verify everything is working normally:

- Glance at the top graph—traffic should look normal for the time of day
- Check device count—does it match what you expect?
- Look at alarm count—zero or low is good

# Chapter 4: Managing Devices on Your Network

Device management is where theory becomes practical. This is where you identify what's on your network, organize it logically, and apply policies to control behavior.

This chapter walks through everything related to devices: viewing them, organizing them, monitoring their activity, and controlling their access.

## The Device List

The device list is the inventory of everything connected to your network. Access it by tapping the Devices button on the dashboard.

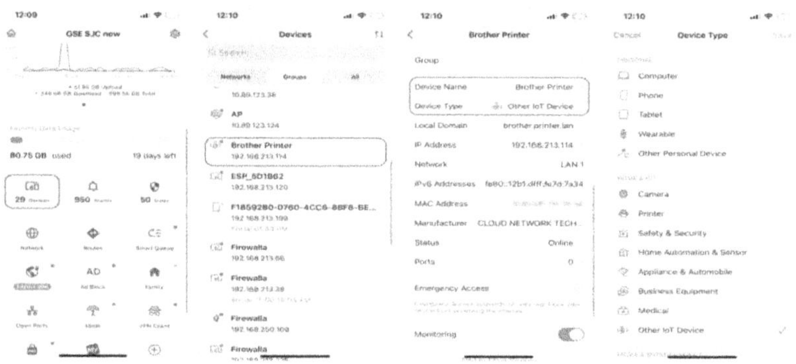


## Device List

Tap any device to open its detail page. This is where you see everything about that device and configure how it's managed.

## Device Groups

Device groups let you manage multiple devices as a unit. Instead of creating the same rule for ten different devices, create one group containing those devices and apply the rule to the group.

## Users

Users are an enhanced version of groups designed specifically for people. Where groups are device-focused, users are person-focused.

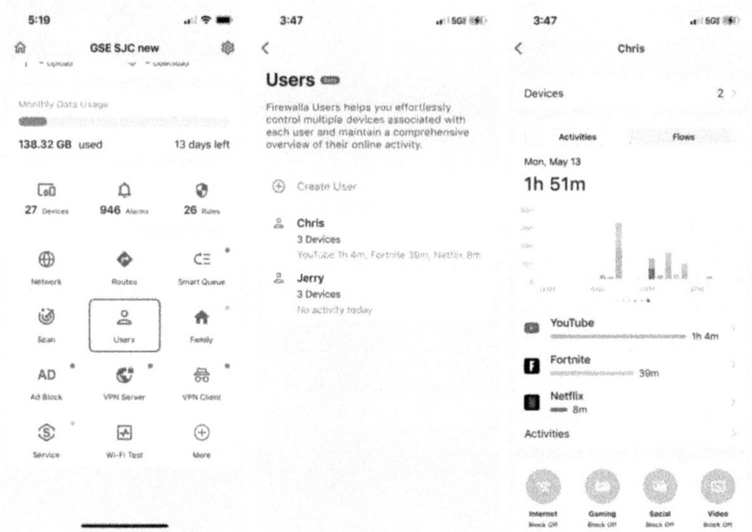


*User activity tracking shows detailed time spent on different apps and services across all devices.*

# Chapter 5: Essential Security Features

This is where Firewalla proves its value. The features in this chapter are what protect you from threats, block intrusions, and prevent malware from reaching your devices.

We'll walk through each major security feature, explain how it works, when to use it, and how to configure it for your needs.

## Active Protect

Active Protect is Firewalla's intrusion detection and prevention system. It's the core of your network security, running continuously to identify and block threats.

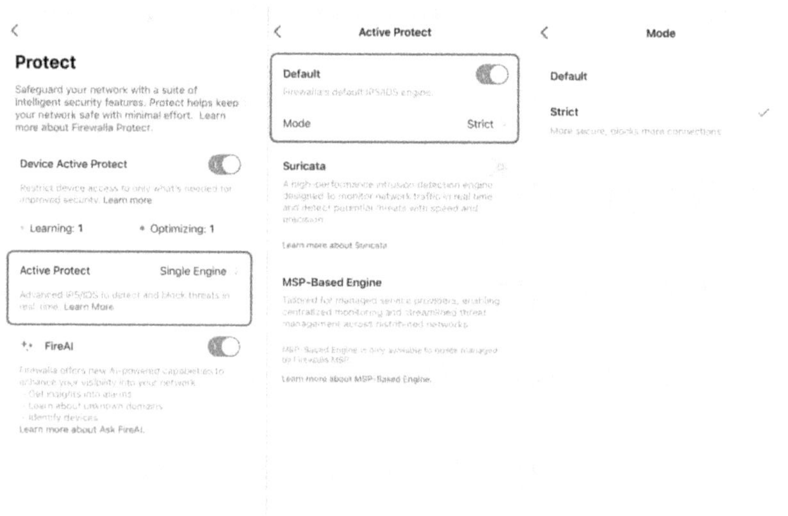


*Active Protect settings allow you to adjust the protection level for your network.*

## Ad and Tracker Blocking

Ad blocking happens at the DNS level. When a device tries to load an ad or tracker, it asks DNS to resolve the

domain name. Firewalla intercepts this request and refuses to answer for known advertising and tracking domains.

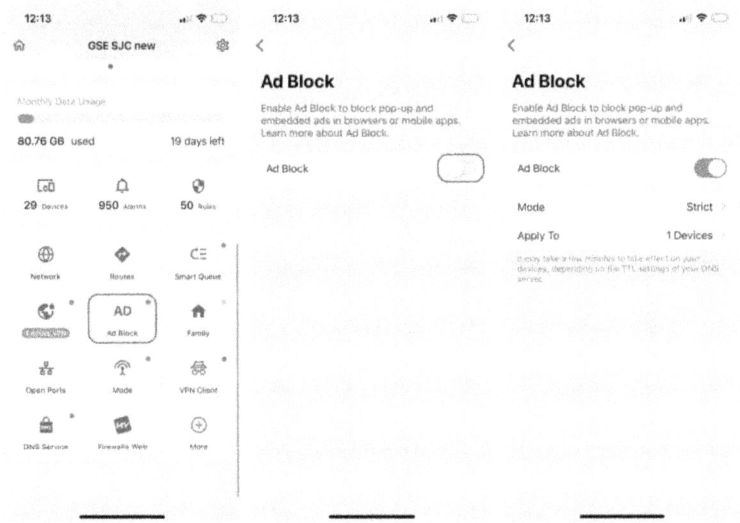


*Ad Block settings allow you to control what gets blocked and how aggressively.*

# Geo-IP Filtering

Geo-IP filtering blocks traffic based on geographic location. You can block all traffic to or from specific countries.

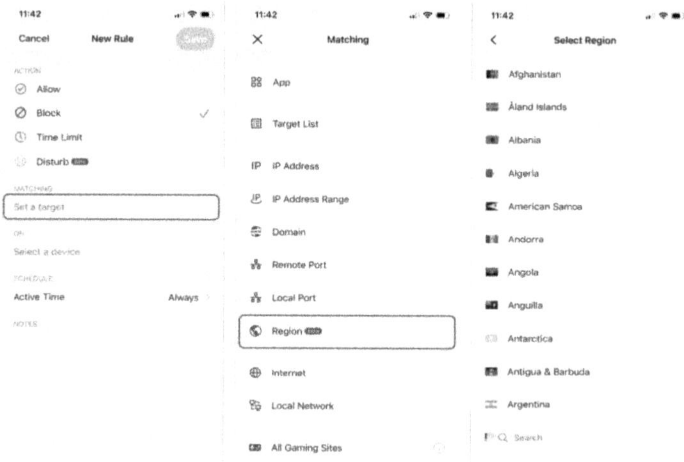


*Geo-IP filtering lets you block traffic based on geographic location.*

# Security Scans

Firewalla can scan your network for vulnerabilities and security issues. These scans are manual—you initiate them when you want to check your network's security posture.

# Chapter 6: Family Protection

Managing internet access for children is one of Firewalla's most-used capabilities. This chapter covers everything from content filtering to time limits to activity monitoring.

The goal isn't to spy on or restrict children unnecessarily—it's to create appropriate boundaries that protect them while giving them the freedom to learn and explore online safely.

## Family Protect

Family Protect is Firewalla's content filtering feature. It blocks access to adult content, violent material, and other inappropriate sites.

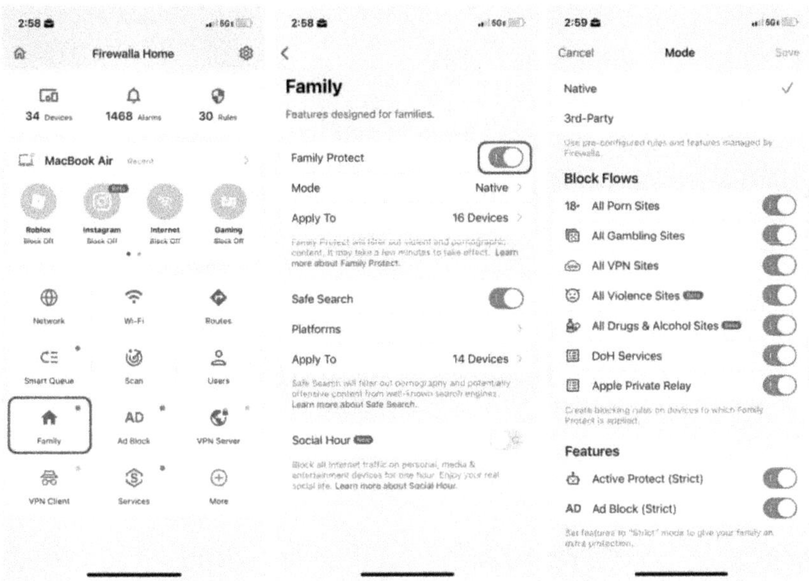

## Safe Search

Safe Search forces search engines to filter their results, removing explicit content from search results. This works with Google, Bing, DuckDuckGo, and YouTube.

## Time Schedules and Rules

For more sophisticated control, create time-based rules that automatically enforce boundaries.

## Activity Monitoring

If you've created Users for family members, you get detailed activity tracking automatically.

## Creating Effective Boundaries

Some philosophy about using these features effectively:

### Communication First

Don't deploy parental controls secretly. Children should know what rules are in place and why.

# Chapter 7: VPN Features

Virtual Private Networks (VPNs) are one of Firewalla's most powerful capabilities. They provide encrypted tunnels for secure remote access, protect your privacy when traveling, and enable site-to-site connections between locations.

This chapter covers everything VPN-related: running your own VPN server for remote access, connecting as a client to third-party VPN services, and linking multiple networks together.

## VPN Server

Firewalla's VPN server lets you connect back to your network from anywhere. When you're traveling, at a coffee shop, or on public Wi-Fi, you can tunnel through your home network for security and access to local resources.

### Why Run Your Own VPN Server

Commercial VPN services protect you from your ISP, but they see all your traffic instead. Running your own VPN server gives you:

- Complete privacy: Only you control the server.
- Access to home resources: Connect to your NAS, printers, cameras.
- Home network protection: All your Firewalla security features apply even when traveling.
- Trusted IP address: Appear to be browsing from your home IP.
- No monthly fees: Unlike commercial VPNs, this costs nothing beyond internet service.

### OpenVPN vs WireGuard

Firewalla supports two VPN protocols. Both work well, but they have different characteristics.

### OpenVPN

OpenVPN is the older, more established protocol. It's been around since 2001.

- Universal client support: Every platform has mature OpenVPN clients.
- Can run over TCP or UDP: TCP mode works through more restrictive firewalls.
- Well-tested: Decades of use have proven reliability.

Disadvantages: Slower performance, higher battery drain on mobile.

### WireGuard

WireGuard is the newer protocol, designed for modern cryptography and performance.

- Much faster: 1.5-2x the throughput of OpenVPN.
- Lower battery usage: More efficient on mobile devices.
- Simpler code: Easier security auditing.
- Modern cryptography: Uses current best-practice algorithms.

**Design considerations:** Operates over UDP and offers less configuration flexibility than OpenVPN.

## Setting Up VPN Server

From the main screen, tap VPN Server. Choose OpenVPN or WireGuard, then enable it.

Firewalla automatically generates keys, configures the server, and attempts to set up port forwarding.

## VPN Client

VPN Client feature lets you route traffic from your network through third-party VPN services. This is useful when you want privacy from your ISP or need to access region-restricted content.

### When to Use VPN Client

- Privacy from ISP: Hide your browsing from your internet provider.
- Access restricted content: Access services limited by geographic location.
- Work requirements: Company VPN for remote access.
- IoT through VPN: Route smart TV or game console through VPN without installing client on device.

### Configuring VPN Client

Go to VPN Client -> Create Profile -> 3rd Party VPN. Import configuration file from your VPN provider.

Major providers supported: NordVPN, ExpressVPN, ProtonVPN, Private Internet Access, and others. Each provides configuration files you can download.

# Chapter 8: Advanced Networking

This chapter covers advanced networking features for fine-grained control. These features are primarily for Gold and Purple users.

## Network Segmentation

Network segmentation divides your network into separate sections with controlled communication between them.

### Why Segment?

- IoT isolation: Keep smart devices separated from computers.
- Guest network: Give visitors internet without exposing private devices.
- Work separation: Isolate work devices from personal network.
- Performance: Reduce broadcast traffic.

### Port-Based Segmentation

Use Firewalla's physical ethernet ports. Each port can be its own network with unique IP range.

Example: Port 1 - Main (192.168.1.x), Port 2 - IoT (192.168.2.x), Port 3 - Guest (192.168.3.x)

## VLANs

VLANs let you create multiple virtual networks on a single physical port. This overcomes port limitations.

### Setting Up VLANs

Network Manager -> Create Network -> Local Network

- Name: Descriptive name

- Type: VLAN
- VLAN ID: Choose a number (10, 20, 30)
- Port: Select physical port
- IP Range: Unique subnet

## Multi-WAN

Multi-WAN uses two internet connections for redundancy or additional bandwidth.

### Failover Mode

One primary, one backup. All traffic uses primary until it fails, then automatically switches to backup.

### Load Balancing Mode

Both connections actively carry traffic. Firewalla distributes to maximize total bandwidth.

# Chapter 9: Troubleshooting and Maintenance

This chapter teaches systematic troubleshooting, diagnostic interpretation, and system maintenance.

## Common Issues and Solutions

### Internet Connection Problems

#### Internet Down Event

- Check physical connections: Cable from modem to Firewalla secure?
- Check modem status: Lights indicating normal operation?
- Verify WAN connection: Network Manager shows IP address?
- Test bypass: Computer direct to modem works?

### VPN Connection Issues

#### VPN Client Cannot Reach Server

- Verify server is running: Check VPN Server page shows Active
- Check port forwarding: VPN port correctly forwarded?
- Test from external network: Use cellular not home Wi-Fi
- Verify DDNS: Hostname resolves to current IP?

## Network Events

Firewalla logs significant connectivity changes as Network Events. Access through Network Performance widget or Settings -> Events.

### Interpreting Events

### High Latency Detected

Indicates slow response times. Usually means:

- ISP network congestion
- Problem with internet infrastructure
- Bandwidth saturation on your connection

### High Packet Loss

Data packets not reaching destination. Causes:

- Poor Wi-Fi signal quality
- Failing network hardware
- ISP network issues

# Chapter 10: Network Monitoring and Optimization

Understanding network performance is essential for maintaining good user experience. This chapter covers monitoring and optimization features.

## Network Performance Monitoring

Firewalla continuously monitors key metrics for network health.

### The Network Performance Widget

Color-coded bar chart provides at-a-glance status:

- Green: Normal performance
- Yellow: Degraded (high latency or packet loss)
- Red: Complete outage

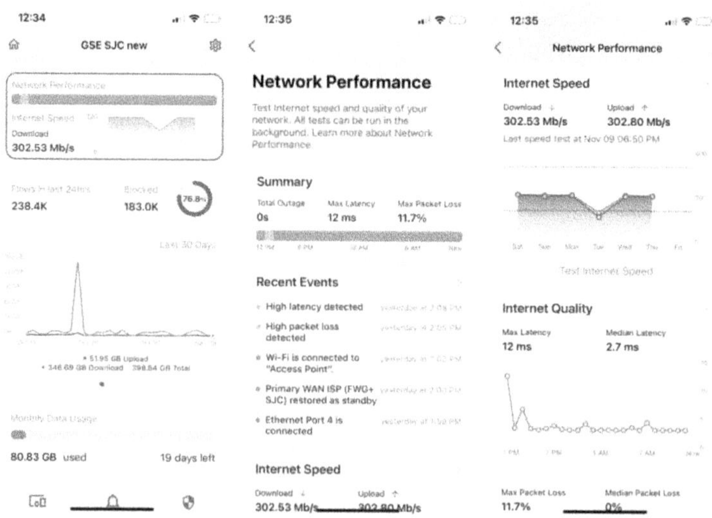


*Network Performance widget provides immediate visibility into network status.*

## Speed Testing

Automated and manual speed tests track internet performance over time.

### Automatic Testing

Network Performance -> Internet Speed -> Test Options

- Frequency: Daily, 12 hours, or custom
- Test server: Auto or specific
- Time of day: Off-peak for accurate baseline

## Bandwidth Monitoring

Track who uses bandwidth and how much.

### Live Throughput

Real-time graph shows current bandwidth usage. Perfect for identifying bandwidth hogs in the moment.

### Historical Usage

Data Usage feature (More -> Data Usage):

- Monthly view: Total usage and trends
- Daily breakdown: Drill down to high-usage days
- Per-device: Identify bandwidth hogs
- Top destinations: Which sites receive most traffic

## Smart Queue

Smart Queue prioritizes important traffic during congestion. Gaming and video calls get priority over downloads.

### Enabling Smart Queue

Network -> LAN -> Edit -> Smart Queue. Enable and set speeds (use 90-95% of actual speeds).

## Network Optimization Tips

### For Gaming

- Enable Smart Queue with accurate speeds
- Use wired connection
- Create QoS rule prioritizing gaming

### For Video Calls

- Enable Smart Queue
- Pause non-essential downloads during calls
- Use wired connection when possible

### For Streaming

- Enable Smart Queue for multiple streams
- Check bandwidth usage during buffering
- Limit less important devices during peak viewing

# Chapter 11: Advanced Rules and Policies

Rules are the primary way you control traffic on your network. While basic rules are straightforward (block this site, allow that device), advanced rules combine multiple conditions, use schedules, and leverage sophisticated targeting to create nuanced policies.

This chapter explores advanced rule concepts: target lists, policy-based routing, smart queue rules, and complex scheduling. You'll learn to build sophisticated policies that adapt to different situations automatically.

## Understanding Rule Priority

When multiple rules could apply to the same traffic, Firewalla needs to know which rule wins. Understanding priority is essential for creating effective policies.

### The Priority Hierarchy

Rules are evaluated in this order:

- Device-specific rules: Rules targeting individual devices have highest priority
- Group rules: Rules targeting device groups come next
- Network rules: Rules for entire networks (VLANs, LANs)
- Global rules: Rules applying to all devices have lowest priority

Within each level, rules are checked in the order they appear in your rules list. The first matching rule determines what happens to the traffic.

### Example Scenario

You have these rules:

- Global rule: Block all video sites
- Group rule for Kids group: Allow YouTube
- Device rule for Sarah's iPad: Block YouTube

Result: Sarah's iPad blocks YouTube (device rule wins), other kids can access YouTube (group rule wins over global), adults have all video sites blocked (global rule applies).

## Target Lists

Target lists let you group multiple domains or IP addresses into a single reusable list. This simplifies rule management and makes policies more maintainable.

### Creating Target Lists

Target lists are managed through the Firewalla web interface (not the mobile app). Go to my.firewalla.com, log in, and select Target Lists from the sidebar.

To create a new list:

- Click Create New List
- Name it descriptively (Work Domains, Gaming Servers, Social Media)
- Add entries: domains (exact or wildcard) or IP addresses (single or range)
- Save the list

### Domain Entries

You can specify domains in several ways:

- Exact domain: example.com (matches only that exact domain)

- Wildcard: *.example.com (matches all subdomains like www.example.com, api.example.com)
- Both: Create two entries if you want example.com AND all subdomains

**IP Entries**

- Single IP: 192.168.1.50
- IP range in CIDR: 192.168.1.0/24 (entire subnet)
- Custom range: 192.168.1.100-192.168.1.200

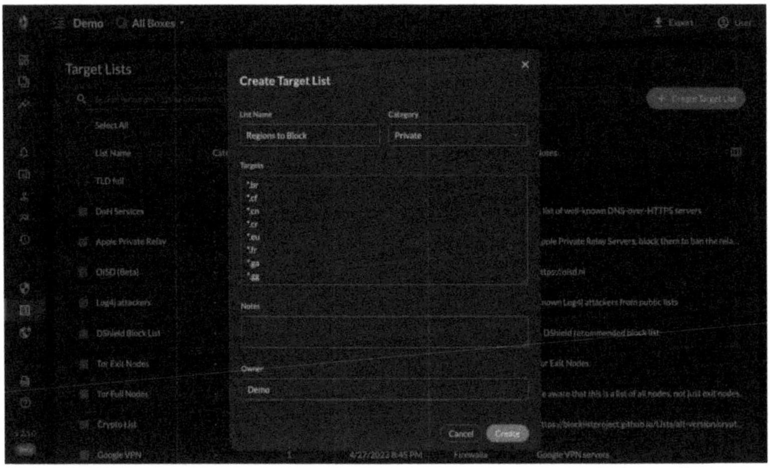


*Target Lists are created and managed through the web interface.*

## Using Target Lists

Once created, target lists appear as options when creating rules:

- Block Rule: Block access to all entries in the list
- Allow Rule: Permit access to list entries (useful for exceptions)
- Route Rule: Send traffic to list destinations through specific interface

- Smart Queue Rule: Prioritize or limit bandwidth for list traffic

**Common Target List Uses**

- Work domains: List of company sites that should route through work VPN
- Streaming services: All major streaming sites for bandwidth limiting
- Social media: Facebook, Instagram, Twitter, TikTok, Snapchat for time limits
- Gaming: Game servers for prioritization or blocking
- Ad/tracker domains: Additional blocking beyond built-in lists

## Policy-Based Routing

Policy-based routing (PBR) sends different traffic through different network paths based on rules you define. This is more sophisticated than traditional routing which only considers destination IP.

### Common PBR Scenarios

**Work VPN for Specific Domains**

You need to access company resources through work VPN, but don't want all traffic going through it:

- Create target list with work domains
- Create route rule: Target List = Work Domains, Interface = Work VPN, Device = Work Laptop
- Result: Only work sites go through VPN, everything else uses regular internet

**Streaming Through VPN**

Access region-locked streaming content:

- Create route: Target = Netflix (or target list with streaming services)
- Interface = VPN connection in desired region
- Device = Smart TV or specific devices

### Multi-WAN Traffic Splitting

Use different internet connections for different purposes:

- Video conferencing through fast, low-latency connection
- Bulk downloads through slower but unlimited connection
- Gaming through connection with best latency

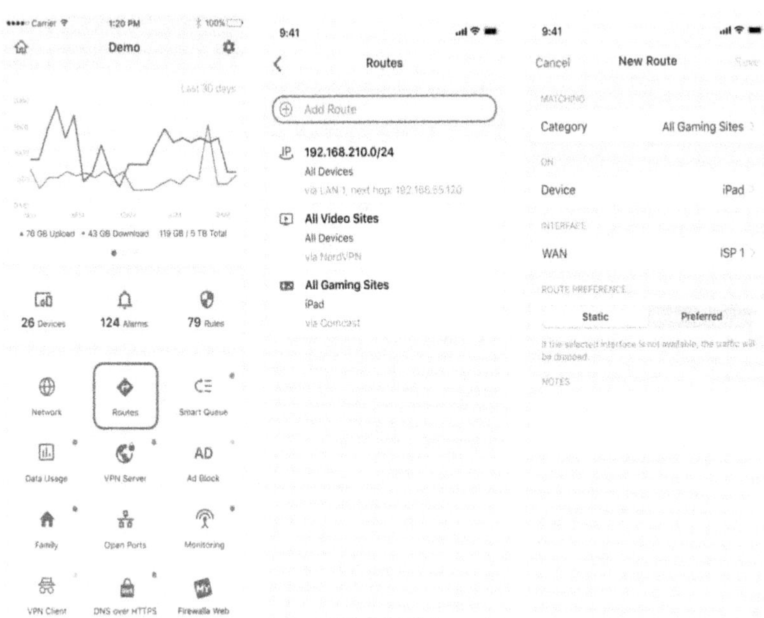


*Route rules direct traffic through specific network interfaces based on your criteria.*

## Smart Queue Rules

Smart Queue manages bandwidth allocation and prioritization. While basic Smart Queue works automatically, custom rules give you fine control.

### Creating Smart Queue Rules

Go to Smart Queue -> Smart Queue Rules -> Add Rule

Configure:

- Target: What traffic this rule applies to (domain, app, category, target list)
- Device/Group: Which devices this affects
- Action: Priority (High/Normal/Low) or Bandwidth Limit
- Schedule: When the rule is active (optional)

#### Priority Levels

-

- High: Video calls, gaming, VoIP - get bandwidth first, minimal delay
- Normal: Most traffic - fair sharing
- Low: Downloads, updates - use leftover bandwidth

#### Bandwidth Limiting

Instead of prioritization, you can cap bandwidth:

- Limit streaming to 5 Mbps to ensure other activities don't suffer
- Cap background downloads to 1 Mbps
- Restrict guest network to 10 Mbps total

## Scheduled Rules

Rules can activate and deactivate automatically based on time. This creates dynamic policies that adapt to your routine.

## Schedule Types

### Time of Day

Activate rules during specific hours:

- Block games 9 AM - 3 PM on school days
- Limit streaming midnight - 6 AM to preserve bandwidth for security cameras
- Enable strict content filtering during kids' internet time

### Days of Week

- Weekday rules: Strict during school/work week
- Weekend rules: Relaxed policies on Saturday/Sunday

### Complex Schedules

Combine multiple time blocks:

- School nights: Limit to 1 hour, 7-8 PM only
- Weekend mornings: No restrictions until noon
- Weekday afternoons after school: 2 hours, homework sites allowed

# Chapter 12: Advanced Security Topics

This chapter dives deeper into security features, covering advanced configurations, threat detection, security best practices, and how to respond to different types of security events.

## Defense in Depth

Security isn't one feature—it's multiple layers working together. If attackers bypass one layer, others catch them.

### Security Layers in Firewalla

- Network perimeter: Firewall blocks uninvited external connections
- Active Protect: IDS/IPS detects and blocks known attack patterns
- DNS filtering: Blocks malicious domains before connections start
- Device isolation: Segmentation limits damage if one device is compromised
- Access control: Rules restrict what devices can do
- Monitoring: Alerts notify you of suspicious activity

## Advanced Active Protect

We covered Active Protect basics earlier. Here we explore advanced configurations and understanding what it's protecting against.

### Strict Mode vs Default Mode

Active Protect has two sensitivity levels:

#### Default Mode

Balanced protection with minimal false positives. Blocks:

- Known exploits and attacks
- Malware connections
- Clear intrusion attempts
- Obviously malicious traffic

Rarely interferes with legitimate services.

**Strict Mode**

Maximum protection, may cause some false positives. Additionally blocks:

- Suspicious patterns that might be attacks
- Port scans and reconnaissance activity
- Connections to newly registered domains
- Traffic patterns associated with malware (even if not definitively malicious)

Use Strict mode if:

- Security is paramount (business, sensitive data)
- You're willing to troubleshoot occasional false positives
- Network handles valuable information

## Understanding Security Alarms

Security alarms require different responses depending on their type and severity.

### Alarm Types and Responses

**Malware Connection Blocked**

A device tried to connect to known malware servers.

Response:

- Identify which device triggered alarm
- Check if device has been acting strangely (slow, popup ads, battery drain)
- Run antivirus scan on that device
- Consider reinstalling OS if seriously compromised
- Change passwords used on that device

**Intrusion Attempt**

Someone from outside tried to exploit your network.

Response:

- This is usually normal internet background noise - automated bots scanning for vulnerabilities
- Firewalla blocked it, so you're protected
- No action needed unless you see many attempts from same source
- If repeated attacks from same IP, create block rule for that IP or country

**Abnormal Behavior**

Device is behaving unusually (strange connections, high bandwidth, odd timing).

Response:

- Review device's recent activity in flows
- Check if behavior matches legitimate use (large upload might be cloud backup)
- Investigate if behavior is unexplained
- Consider device compromise if clearly abnormal

## Zero Trust Principles

Zero trust means trust nothing by default. Every device and connection must prove it's legitimate.

## Implementing Zero Trust

- New device quarantine: All new devices start isolated until you verify and approve them
- Network segmentation: Devices can't see other segments by default
- Least privilege: IoT devices can only access their cloud services, nothing else on your network
-
- Continuous monitoring: Even trusted devices are watched for suspicious behavior

## Practical Example

Smart TV with zero trust:

- Segment: IoT network, can't see computers or NAS
- Rules: Can only connect to manufacturer's servers, all other destinations blocked
- Monitoring: Alerts if it tries to connect to unexpected destinations

If TV is compromised, attacker gains nothing—can't reach your valuable devices or data.

# Chapter 13: Automation and Integration

Firewalla can integrate with other systems and automate responses to events. This chapter covers webhooks, IFTTT integration, API access, and building automated workflows.

## Webhooks

Webhooks send HTTP requests when events occur. This lets external systems react to Firewalla events.

### Setting Up Webhooks

Settings -> Advanced -> Webhooks

Enable webhooks and configure:

- URL: Where to send notifications
- Events: Which events trigger webhooks (alarms, device connections, etc.)
- Authentication: Optional secret token for security

#### What You Can Do

- Send notifications to Slack, Discord, or Microsoft Teams
- Log events to external database
- Trigger automation in home automation system
- Integrate with security monitoring systems

## IFTTT Integration

IFTTT (If This Then That) connects Firewalla to hundreds of services and devices.

#### Example Automations

- If new device connects, send text message

- If security alarm triggered, flash smart lights red

- If kid's device goes online after bedtime, send notification
- If bandwidth exceeds threshold, log to Google Sheets

## API Access

For developers, Firewalla provides API access for custom integrations.

### What the API Enables

- Query device list and status programmatically
- Create/modify rules via scripts
- Pull bandwidth statistics for custom reporting
- Integrate with custom dashboards

## Home Automation Integration

Firewalla can work with home automation platforms.

### Example Scenarios

- When movie mode activates, create rule prioritizing streaming, blocking notifications
- When vacation mode starts, enable strict security settings
- When sleep mode begins, pause kids' devices automatically
- When security alarm triggers, activate security cameras recording

# Chapter 14: Maintenance and Updates

Like any network device, Firewalla requires occasional maintenance. This chapter covers updates, backups, performance optimization, and long-term care.

## Software Updates

Firewalla releases updates regularly with new features, security patches, and bug fixes.

### Update Types

#### Automatic Updates

By default, Firewalla automatically updates during off-peak hours (typically 2-4 AM local time).

Updates are:

- Seamless: No downtime, connections stay active
- Tested: All updates go through beta testing before release
- Reversible: Can roll back if issues occur

#### Manual Updates

You can manually update anytime:

- Settings -> System -> Software Update
- Check for Updates
- Install Now if update is available

#### Beta Channel

Advanced users can opt into beta updates to test new features early:

- Settings -> System -> Update Channel -> Beta

Beta updates may have bugs. Only recommended if you're comfortable troubleshooting.

## Backups

Regular backups protect your configuration.

### What Gets Backed Up

- All rules and policies
- Device names and groups
- Network configuration
- VPN settings
- Security settings

#### What Does NOT Get Backed Up

- Historical data (bandwidth usage, flows)
- Logs and alarms
- Temporary state

### Creating Backups

Settings -> Backup & Restore -> Create Backup

Backup downloads to your device. Store it safely:

- Cloud storage (Google Drive, Dropbox)
- Local NAS or file server
- External hard drive

Create backups:

- Before major configuration changes
- After setting up complex rules
- Monthly for peace of mind

### Restoring from Backup

Settings -> Backup & Restore -> Restore

Select backup file. Firewalla restores configuration and restarts.

After restore:

- Verify network connectivity
- Check that devices reconnect properly
- Test key features (VPN, specific rules)

## Performance Monitoring

Monitor Firewalla's performance to ensure it's handling your network well.

### Key Metrics

- CPU usage: Should average under 50 percent during normal use
- Memory usage: Monitor trends, shouldn't constantly increase
- Temperature: Firewalla should stay cool, ensure adequate ventilation
- Connection count: Track as network grows

#### When to Upgrade Hardware

Consider upgrading if:

- CPU consistently over 70 percent
- Memory constantly maxed
- Network speed slower than internet connection capability
- Growing beyond device limits (Blue -> Gold, Gold -> Purple)

# Routine Maintenance

## Monthly Tasks

- Review alarms: Check for patterns or recurring issues
- Update device names: Rename unknown devices you've identified
- Review rules: Delete unused rules, update outdated ones
- Check for updates: Ensure you're on current version

## Quarterly Tasks

- Create backup
- Review security settings: Are they still appropriate?
- Audit device list: Remove old devices no longer used
- Review bandwidth patterns: Any unexpected changes?

## Annual Tasks

- Complete security audit: Review all settings
- Update VPN profiles: Regenerate keys for security
- Review network segmentation: Still makes sense?
- Performance evaluation: Is hardware still adequate?

# Appendix A: Quick Start Guide

This condensed guide gets you from box to working network in minimum time.

## Initial Setup (15 minutes)

- Connect WAN port to modem
- Power on Firewalla Gold
- Download Firewalla app on phone
- Follow in-app setup wizard
- Choose Router Mode
- Set administrator password
- Connect devices to Firewalla's network

## Essential Configuration (30 minutes)

- Name your devices (Devices -> tap each device -> set name)
- Enable Active Protect (Settings -> Active Protect -> On)
- Enable ad blocking (Settings -> Ad Block -> Normal)
- Set up one security alarm notification

## Recommended Next Steps

- Create device groups (family members, IoT, guests)
- Set up VPN server for remote access
- Configure parental controls if needed
- Review and acknowledge alarms daily for first week

# Appendix B: Troubleshooting Common Issues

## Internet Not Working

### Symptoms

Devices show connected to Wi-Fi but can't access internet

### Solutions

- Check WAN cable connection
- Reboot modem, wait 2 minutes, then reboot Firewalla
- Verify WAN shows IP address in Network Manager
- Test bypassing Firewalla (computer direct to modem)

## Slow Internet

### Symptoms

Web pages load slowly, videos buffer

### Solutions

- Check bandwidth graph for congestion
- Identify device using excessive bandwidth
- Run speed test - is ISP delivering promised speed?
- Enable Smart Queue if not already active

## VPN Not Connecting

### Symptoms

VPN client shows error or connection timeout

### Solutions

- Verify VPN server shows Active in app
- Check port forwarding is configured

- Test from external network (cellular data)
- Verify DDNS resolves to current public IP

## Device Cannot See Other Devices

### Symptoms

Can't print, can't access NAS, devices can't discover each other

### Solutions

- Check if devices are on different networks
- Review rules blocking local network access
- Disable VqLAN if enabled on one of the devices
- Create allow rule for specific device-to-device connectionAppendix C: Glossary

# Appendix C: Glossary

### Active Protect

Intrusion detection and prevention system that identifies and blocks known attacks and exploits.

### DHCP

Dynamic Host Configuration Protocol. Automatically assigns IP addresses to devices on your network.

### DNS

Domain Name System. Translates domain names (google.com) into IP addresses (142.250.185.46).

### IDS/IPS

Intrusion Detection System / Intrusion Prevention System. Monitors network for malicious activity and can block threats.

### LAN

Local Area Network. Your private network - devices connected to your Firewalla.

### MAC Address

Media Access Control address. Unique hardware identifier for each network device.

### NAT

Network Address Translation. Allows multiple devices to share one public IP address.

### Port Forwarding

Directs external connections to specific devices on your network. Required for VPN server, game servers, etc.

**QoS**

Quality of Service. Prioritizes important traffic over less important traffic.

**Router Mode**

Firewalla acts as your main router, handling all network routing and security.

**Smart Queue**

Firewalla's bandwidth management system that reduces congestion and prioritizes traffic.

**Target List**

Collection of domains or IP addresses used as a group in rules.

**VLAN**

Virtual Local Area Network. Creates separate networks on single physical connection.

**VPN**

Virtual Private Network. Encrypted tunnel for secure remote connections.

**WAN**

Wide Area Network. Your connection to the internet (typically your ISP connection).

## Continue the Journey

If this book resonated with you, here's where to go next:

---

**Bitwarden Essentials**
A Practical Guide to Password Security
Secure your passwords with confidence.

**Guard Your Money**
Protect your accounts with security keys.

**Firewalla: After Setup** *(Recommended Next)*
A step-by-step guide to setting up a Firewalla firewall.
Strengthen your network with Firewalla.

---

For readers interested in exploring Awareness,

**Awareness** *(Recommended Next)*
*Noticing What's Already Here*
Begin by noticing the voice inside your head and stepping out of automatic identification with it.

**The Awareness Companion**
*Living with Awareness in Daily Life*
Continue noticing how awareness shows up in real situations, habits, and patterns.

---

*All titles are part of the* **Personal Security in a Digital World** *series.*

**Bitwarden Essentials**

A practical guide to securing and managing your passwords using Bitwarden, with step-by-step instructions for setup, daily use, and account recovery.

**Guard Your Money**

A clear, practical guide to protecting financial accounts using YubiKey hardware security, covering phishing, identity theft, and modern account takeover threats.

**Network Security Essentials**

A plain-language guide to securing a home network using Firewalla Gold Plus, focused on understanding security choices rather than just technical settings.

## Firewalla: After Setup

*Firewalla: After Setup* focuses on the decisions that matter once your network is running: how to structure devices, reduce noise, contain risk, and keep security effective over time — without constant tweaking.

# The Awareness Series

These books explore awareness as it is experienced — not as a concept, but as something noticed in real time.

***Awareness***

*Noticing the Voice Inside Your Head*
The starting point — recognizing the voice and stepping out of automatic identification with it.

**The Awareness Companion**
*Living with Awareness in Daily Life*
Continue noticing how awareness shows up in real situations, habits, and patterns.

The continuation — living with awareness over time, including the parts that are often left out.

*These books can be read in any order, but they follow a natural progression:*

*Awareness → The Awareness Companion*

## Also Available from Prairie Light Press

The **Awareness Series** explores the nature of thought, attention, and the sense of self—through direct observation, not theory.

Written in clear, simple language, these books are designed to be experienced, not analyzed.

Available worldwide in Kindle and paperback editions.

Visit:

Prairie Light Press
https://www.prairielightpress.com

## Also Available from Prairie Light Press

The ***Personal Security in a Digital World*** series helps readers protect passwords, financial accounts, and home networks using practical, plain-language guidance.

Available worldwide in Kindle and paperback editions.

Visit:

**Prairie Light Press**

http://www.prairielightpress.com/

---

## Prairie Light Dispatch

For more reflections on awareness, life, and technology, visit *Prairie Light Dispatch* on Substack at

**Prairie Light Dispatch**

https://lgphil7.substack.com/

Continue Reading for an

Exclusive Prairie Light Press Reader Preview

Of

**Firewalla**

*After Setup*

**Continue Your Journey**

You've taken an important step toward protecting your digital life by securing your passwords, financial and other important accounts using a YubiKey, and you installed a hardware-firewall.

At this point, you have multiple layers doing their specific work protecting you.

This book is for users of Firewalla Gold Plus after the installation. This book guides you through setting up Active Protect, Intrusion Detection, and many of the features available in the Firewalla.

Prairie Light Press publishes books on technology, security, financial protection, and personal awareness—all designed to help you make better decisions in an increasingly complex world.

The following pages contain a preview of another Prairie Light Press title.

# Book Preview

## Firewalla

*After Setup*

Turn the page to enjoy an exclusive preview from Prairie Light Press.

Exclusive Preview from Prairie Light Press

# Firewalla

## *After Setup*

An Operator's Guide to
Running a Secure Home Network

L Phillips

PRAIRIE LIGHT PRESS
ESTD. 2025

*Helping readers move from the unknown to the known*

## Table of Contents



Get Started Right Away with our Quick-Start Guide on page 5

## QUICK-START FOR OPERATORS (90MINUTES)

This book is designed to be read end-to-end, but you don't need to do everything at once.

If you have about **90 minutes**, this section gives you the highest-security return for your time. You can come back later for depth.

---

## 0–10 MINUTES: CLARITY

- Read **Chapter 1** to adopt the operator mindset
- Skim **Chapter 2** to understand device discovery
- Ensure most devices are identified and named (perfection not required)

Goal: You know what's on your network and why.

---

## 10–30 MINUTES: STRUCTURE

- Read **Chapter 3** (Naming Conventions)
- Read **Chapter 4** (Grouping Strategy)

**Apply immediately:**

- Use a consistent naming convention
- Create at least **Trusted**, **IoT**, and **Guest** groups
- Assign every device to a group

Goal: You can reason about your network at a system level.

---

## 30–45 MINUTES: BASELINE PROTECTION

- Read **Chapter 5** (Built-In Protections)**Enable:**
- Active Protect (Default mode)
- Ad Block (network-wide or IoT-only)

Goal: To remove ambient risk without ongoing effort.

---

## 45–70 MINUTES: CONTAINMENT

- Read **Chapter 6** (One-Way Communication)
- Read **Chapter 7** (Blocking Lateral Movement)

**Apply:**

- Block IoT devices from initiating connections to local networks
- Block IoT-to-IoT communication

Goal: Prevent small compromises from spreading.

---

## 70–90 MINUTES: EXPOSURE REDUCTION & HABITS

- Skim **Chapter 8** (Region & Country Blocking)
- Read **Chapter 9** (Operational Awareness)

**Apply:**

- Add conservative geo-blocking for IoT or Guest groups
- Establish a simple weekly review habit

Goal: Reduce noise and make problems obvious.

## WHAT TO SKIP FOR NOW

- Fine-grained exceptions
- Advanced tuning
- Edge-case optimizations

Those come later, once the system is stable.

---

## QUICK-START TAKEAWAY

If you complete this path, your network will be:

- Structured
- Contained
- Quiet
- Sustainable

You can now read the rest of the book for depth instead of urgency.

## NOTE TO READER

This edition has been refined for clarity, pacing, and long-term usability.

No concepts were simplified. No security posture was weakened.

You'll notice:

- Cleaner sectioning for scan-reading

- Consistent terminology and tone
- Priority callouts where decision overload is common
- Operator Takeaways that reinforce judgment, not checklists

This is an operational manual. Read actively.

**Disclaimer**

This book provides general guidance based on common home-network environments and typical Firewalla configurations. Every network is different.

You are responsible for understanding your own environment, testing changes incrementally, and adjusting rules to fit your specific needs.

No security configuration is risk-free, and no guidance replaces informed judgment.

# INTRODUCTION — BEYOND BASIC SETUP

So you've got Firewalla installed. It's running. The lights are blinking. Devices are connecting. Internet works.

Congratulations—you've completed the easy part.

Now comes the real work: actually using Firewalla effectively. This is where most people struggle. Not because Firewalla is difficult, but because network management requires ongoing attention, decision-making, and refinement. It's not a set-it-and-forget-it appliance.

This book is about that operational phase—the weeks, months, and years after setup. It's about the daily decisions, the maintenance routines, and the judgment calls that determine whether Firewalla becomes an essential security tool or just another box gathering dust.

---

## WHAT THIS BOOK IS (AND ISN'T)

This is not a setup guide. We assume Firewalla is already running in Router Mode with devices connected. Installation, onboarding, and feature tours are deliberately excluded.

This is an operational manual.

Specifically, this book focuses on:

- Making naming decisions that scale
- Organizing devices for long-term management
- Creating rules that survive real-world use

- Recognizing what is normal vs what deserves attention
- Building habits that keep your network secure without constant babysitting

Think of it this way: the setup guide gave you a working car. This book teaches you how to drive it well, maintain it properly, and avoid common accidents.

---

## WHO THIS BOOK IS FOR

You're the intended reader if:

Firewalla is installed and working

- Your device list is cluttered or confusing
- You ignore most alarms because there are too many
- You've created rules that don't seem to help
- You want better security but aren't sure what to *do*

You'll get the most value if you're willing to invest a few focused hours up front. Organization and clarity reduce ongoing effort dramatically.

---

## THE OPERATIONAL MINDSET

Effective network management requires a mindset shift.

Think in systems, not one-off fixes. When something goes wrong, ask whether it represents a pattern that deserves a structural response.

Embrace iteration. Your configuration will evolve. That's normal.

Balance security and usability. Over-restriction leads to workarounds that undermine security entirely.

Invest time early to save time later. Naming and grouping may feel tedious, but they pay dividends forever.

Document decisions. Future-you will not remember why a rule exists.

Identification works best as a loop. Observe first, name devices once you're confident, and revisit periodically as behavior and inventory change.

---

## OPERATOR TAKEAWAYS

- Firewalla effectiveness depends on *how it's operated*, not how it's installed
- This book is about judgment, habits, and systems—not features
- You're adopting an operator mindset, not a consumer mindset

**NEXT:** We stop guessing and start identifying what's actually on your network.

# CHAPTER 1 DEVICE DISCOVERY & IDENTIFICATION

Before you can secure your network, you need to know what's on it.

Most networks begin in partial ignorance: dozens of devices, many unnamed, some unknown. This chapter turns that uncertainty into inventory.

---

## IDENTIFICATION IS A PROCESS, NOT A TASK

Device discovery is not something you finish once. New devices appear. Old ones disappear. Patterns emerge over time.

Treat identification as an investigative process rather than a cleanup chore. This mindset prevents frustration and encourages iteration.

Enable Active Protect first, then turn on Ad Block. Family Protect can be added later once structure and visibility are stable.

---

## IF YOU ONLY DO ONE THING

Identify and correctly name at least 90% of your devices.

A mostly-labeled network is exponentially more manageable than a perfect one that never gets finished.

---

## HOW TO: Access Your Device List

1. Open the Firewalla app
2. Tap 'Devices' in bottom navigation

3. Tap arrows icon (top right) for View Options
4. Use search bar to find devices

---

## HOW TO: Identify Unknown Devices

STEP 1: Check Basic Information
1. Tap device
2. Review Manufacturer, Hostname, MAC

STEP 2: Analyze Network Activity
1. Check Network Flows
2. See what services device connects to

STEP 3: MAC Lookup
1. Visit MAC lookup website
2. Enter first 6 characters

STEP 4: Elimination Method
1. Disconnect suspects one at a time
2. Check if unknown goes offline

---

## HOW TO: Rename Devices

1. Tap device
2. Tap name
3. Enter new name
4. Save

## AUTOMATING DISCOVERY WITH NEW DEVICE QUARANTINE

Manually scanning for unknown devices works, but Firewalla offers a powerful automation layer: New Device Quarantine. When enabled, every new device that joins your network is automatically placed into a locked-down Quarantine Group with pre-defined rules that block internet access and local network communication.

This is operationally significant. Instead of discovering unknown devices after they've already been active on your network, quarantine contains them at the moment of arrival. You decide what gets access — not the device.

**Why this matters:** IoT devices often use MAC address randomization (iOS 14+, Android 10+), which causes them to appear as "new" devices each time the address changes. Quarantine catches these automatically, preventing devices from evading your rules by randomizing their identity.

**How quarantine works in practice:** New devices are placed in the Quarantine Group automatically. The group comes with two pre-defined rules: block internet access and block local network access. You can customize these rules, add your own, or treat the Quarantine Group like any other device group. When you've identified and trusted a device, release it from quarantine and assign it to the appropriate group.

**Operator insight:** Let unknown devices run in quarantine for a day or two while you observe their network flows. Devices that only connect to trusted servers (like Amazon AWS or Google Cloud) are more likely to be legitimate IoT devices.

Watch for more Prairie Light Press Reader Previews

www.ingramcontent.com/pod-product-compliance
Lightning Source LLC
LaVergne TN
LVHW051051210926
848486LV00002B/68
* 9 7 9 8 9 9 6 9 8 0 2 3 9 *